Placement Test Guide

Bothell, WA • Chicago, IL • Columbus, OH • New York, NY

mheonline.com

Copyright © 2014 McGraw-Hill Education

Send all inquiries to:
McGraw-Hill Education
8787 Orion Place
Columbus, OH 43240

ISBN: 978-0-02-129426-8
MHID: 0-02-129426-7

Printed in the United States of America.

1 2 3 4 5 6 7 8 9 RHR 17 16 15 14 13

Contents

Number Knowledge Test and Placement Tests Overview

As part of the **Number Worlds** program, the Number Knowledge Test was developed to measure a student's conceptual knowledge of number (number sense). The level Placement Tests were created to determine where each student should begin instruction within the **Number Worlds** curriculum. Together, these tests are a valuable first step in assessing students' intuitive knowledge of number and their preexisting knowledge of the Common Core State Standard (CCSS) math skills associated with their grade level. Students' test results can be used to confirm or adjust their future lessons, as well as inform and differentiate instruction within your classroom. If administered at both the beginning and end of an instructional period, the Number Knowledge Test and Placement Tests may also be used to measure the progress and developmental growth of a student over time.

Goals of the Number Knowledge Test

1. To determine if a student is functioning at, above, or below age/grade level in number knowledge

2. To determine which number concepts the student has mastered, which she is struggling with, and which she still needs to learn

3. To assess a student's progress over the instructional period or academic year

4. To determine which **Number Worlds** level Placement Test to start testing with in order to pinpoint the program level in which each student should begin her instruction

Test Design

The Number Knowledge Test is an oral test that is administered individually to each student and requires oral responses. Precise instruction for administering and scoring each item is included along with nine Visual Arrays that test the solidity of a student's understanding of number sense and decrease the likelihood of guessing at a correct response.

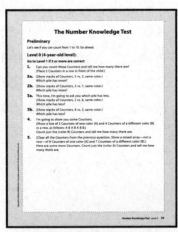

Number Knowledge Test

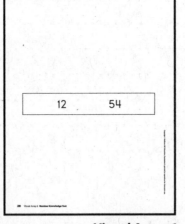

Visual Array 6

Test Record

The Number Knowledge Test Record allows you to record a variety of data about student responses, including the problem-solving strategies used on key items. The raw test score values calculated on the record can be used to determine a student's developmental age score and its grade level equivalent, as well as the corresponding **Number Worlds** program level for instruction.

Goals of the Placement Tests

1. To function as a critical range test in which only items estimated to be within the student's probable range of math understanding are administered

2. To identify in which level a student should begin her instruction within the *Number Worlds* curriculum

3. To assess a student's preexisting knowledge of the Common Core State Standards (CCSS) associated with a level

4. To assess a student's progress over the instructional period or academic year

Placement Tests in Levels A–C

Number Worlds levels A–C are targeted for use by students in grades Pre-Kindergarten through Grade 1.

Placement Tests for levels B and C are designed to be administered orally and individually to each student by a teacher, classroom aid, or parent helper. The tests at these levels consist of teacher's instructions on the left-hand page and reproducible student test masters on the right-hand page.

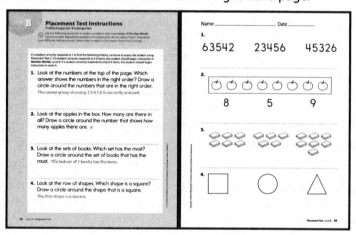

Placement Test, Level B

Placement Tests in Levels D–J

Number Worlds levels D–J are targeted for use by students in Grades 2–8.

At these levels, the Placement Tests consist solely of multiple-choice items. In order to best evaluate the effectiveness of the *Number Worlds* program and prepare the student for future standardized testing, students taking these tests should attempt to take them independently.

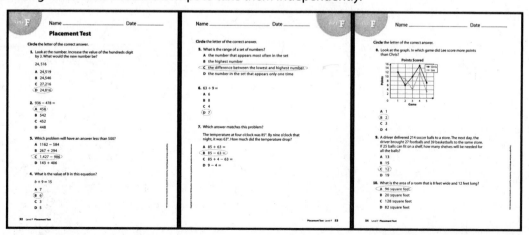

Placement Test, Level F

The Number Knowledge Test

Understanding the Test

Purpose

The Number Knowledge Test was devised to measure the conceptual knowledge of number (number sense) that the average child has available at the age levels of 4, 6, 8, and 10 years. This knowledge has been referred to elsewhere as a set of "central conceptual structures for number" and "powerful organizing schemas" that help children make sense of quantitative problems. Research has shown that the knowledge assessed at each age level of this test is essential for successfully learning arithmetic in school and foundational for higher mathematics learning. A major goal of the **Number Worlds** program is to ensure that all children acquire this knowledge in a well-consolidated fashion at the appropriate age and grade level and have ample opportunity to use it to solve a wide range of quantitative problems.

Design

The Number Knowledge Test is an oral test. It is individually administered to each child and requires oral responses. This feature of the test enables you to assess a child's mental math competencies and conceptual understanding of number. It also enables you to assess the sophistication of a child's problem-solving strategies and to use this information for instructional planning. If, for example, an 8-year-old child consistently counts up from 1 when adding two sets, you can infer that she has not yet acquired the understandings that would enable her to count on from the largest addend to find the sum, so those particular understandings must be carefully nourished and taught. The accompanying Number Knowledge Test Record form is designed to allow you to record responses and problem-solving strategies as you are administering the test.

The Number Knowledge Test is also a developmental test, meaning that knowledge assessed at Level 0 is generally acquired before knowledge assessed at Level 1, and so on. It also means that knowledge at each level of the test is a prerequisite (providing the conceptual building block) for knowledge at the next level of the test. This information is useful for instructional planning. If you know the developmental level of each child in your classroom, you can make informed and appropriate instructional decisions—decisions that enable each child to strengthen her present knowledge and move in easy, manageable steps from one level to the next. The Number Knowledge Test was designed for this purpose.

Age Levels

The labels assigned to each level of this test (e.g., 4-year-old level) actually represent the midpoint in the range of years in which children typically acquire this knowledge. They are meant to suggest that a child's understanding of number changes qualitatively every two years across the middle childhood years, becoming more complex with each change. On average, children acquire the understandings assessed at the 4-year-old level sometime between the ages of 3 and 5 years. The 6-year-old understandings are acquired sometime between the ages of 5–7 years, the 8-year-old understandings between the ages of 7–9 years, and the 10-year-old understandings between the ages of 9–11 years.

Research has shown that children from advantaged backgrounds typically acquire these age-level understandings near the beginning of the age range indicated for each level. Children from less advantaged backgrounds, with fewer learning opportunities, typically acquire these understandings later in the age range. Research has also shown that many children living in low-income communities do not master the understandings expected for their age level (mastery which is typically demonstrated by more advantaged peers) and perform at one or two levels below average on this test. This demonstrates a developmental delay in number knowledge of two or more years.

Developmental Conversion Chart

The test has been standardized based on substantial research data. The following chart correlates Raw Test Scores from The Number Knowledge Test with Chronological Age Equivalents, Grade Level Equivalents, and Number World Program Levels. This correlation indicates the ages and grade levels at which children typically achieve Raw Score values and allows you to assess a child's developmental age in number knowledge understandings based on their score.

This chart can also be used to determine which **Number Worlds** level Placement Test a child should initially be given to pinpoint the level of the program in which they should begin working.

Raw Test Score	Developmental Age Score (Chronological Age Equivalents)	Grade Level Equivalents	Number Worlds Level
1–6	3–4 years	Preschool	Level A
7–8	4–5 years	PreK–K	Level B
9–14	5–6 years	K–1	Level C
15–19	6–7 years	1–2	Level D
20–25	7–8 years	2–3	Level E
26–28	8–9 years	3–4	Level F
29–30	9–10 years	4–5	Level G

Levels of the Test

The Number Knowledge Test is comprised of four levels:

Preliminary: This basic counting item is generally mastered around the age of 3 years. It is included in this test as a warm-up item in order to orient children to the nature of the test and to give them a successful experience at the start. It will also alert you to children who have not yet mastered this skill and who will need opportunities to do so.

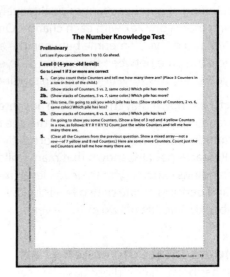

Level 0 (4-year-old level): These items assess a child's ability to count and to quantify small sets when concrete objects (counters) are available and can be touched and manipulated. This knowledge provides an important building block for success at the next level which requires the child to deal with quantities, and changes in quantity, which cannot be touched or seen and have to be imagined.

Level 1 (6-year-old level): There are two classes of items at this level: those which asses a child's knowledge of the number sequence and those that assess a child's ability to handle simple arithmetic problems. Concrete objects are not utilized when administering these items. At this level, a child needs to rely on something like a mental counting line inside her head. Success at this level will give you some idea of whether or not they have constructed this knowledge structure.

Level 2 (8-year-old level): There are two classes of items at this level as well: those which assess knowledge of the number sequence and those that assess knowledge of arithmetic. The primary distinction between items at Level 1 and Level 2 is that items in this level require a child to deal with double-digit numbers (i.e., tens and ones) and/or depend on the use of two mental number lines for successful solution.

Level 3 (10-year-old level): This level also has two classes of items: those which assess knowledge of the number sequence and those that assess knowledge of arithmetic. The primary distinction between items at Level 2 and Level 3 is that items in this level require children to deal with triple-digit numbers (e.g., addition and subtraction problems that require regrouping).

Requirements of the Test

As mentioned, The Number Knowledge Test is an oral test that is administered individually to each child and requires oral responses. Children are not permitted to use paper and pencils to figure out answers. Although this is not an issue at the Kindergarten level, older children often request paper and pencils, guessing (correctly) that the problems would be easier to solve if they did not have to figure them out "inside their heads." To measure conceptual knowledge, however, it is important that children *do* figure things out in their heads.

For example, the Level 2 test question "How much is 12 + 54?" is accompanied by Visual Array 6, which displays those numbers visually to help children remember them.

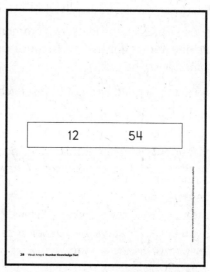

Visual Array 6

If the numbers were stacked vertically in a typical workbook fashion (and if children had a pencil), they could solve this problem easily using single-digit addition. They would simply add the digits in the ones column, record the sum, and then add the digits in the tens column and record the sum. The two sums provide the answer. Many children with a weak understanding of double-digit addition try to solve the problem in this fashion by forming a mental image of the two addends stacked vertically. This process, which imposes a heavy load on memory, is often fraught with errors. By contrast, children who understand double-digit addition can easily solve this problem using base ten understandings, such as adding 10 to 54 to get 64 and then adding 2 more to get 66. Paying attention to the problem-solving strategies children use can tell you much about their level of understanding.

The test has been standardized on the typical United States population, so it is critical in order to obtain a valid and reliable developmental level score that you follow test administration directions precisely. It is always tempting to try to help children solve the problems by rephrasing them or by providing helpful hints. If you do this, however, you will be administering a test much different from the one on which the Developmental Age Scores were based. You may be able to determine how much knowledge the child can demonstrate with assistance, but you will not be able to use the conversion chart to compare the child's performance to her peers across the U.S. This comparison is one of the purposes of this test. Since the Developmental Age Score provides a good starting point for instruction, it is important to administer the test as directed in the following sections.

Administering the Test

Creating a Test Kit

Before actually administering the test, you will need to prepare a test kit with the provided materials. This entails:

☑ collecting 7 Counters of one color and 11 Counters of a different color used in Level 0 items

☑ inserting Visual Arrays 1–9 in protective sleeves and placing them (in order) in a three-ring binder—make sure the top of each array will face the child you are testing who will likely be sitting across from you

☑ inserting a blank page in its own protective sleeve between each array (this allows you to "clean the slate" when one item is finished while concealing the next array from the child until it's called for in the progression of the test)

☑ copying the needed amount of Test Record pages for each child taking the test

It is strongly recommended that you practice giving the test to a friend or colleague before using it with children. This will ensure that test administration and scoring is automatic so that you can focus on the child's responses, identifying her problem-solving strategies, and other behaviors (e.g., expressions of frustration) that provide indications of her level of understanding.

Suggestions

☑ As a general rule, start testing at a level that is at least two years below the age of the child you are testing. This will ensure that children experience success at the beginning level of their test and provide an index of their baseline knowledge.

☑ For children who are 7 or younger, start at the Preliminary Item and continue testing until the child does not pass enough items (amounts indicated on the Test Record) to allow them to progress to the next level.

☑ For children who are 8 or older, you can omit the Preliminary Item and Level 0 and start testing at Level 1. Give the child 5 points for Level 0.

☑ Plan a time when the other children in your class are occupied, so you will not be interrupted.

☑ Choose or edit your location carefully so that materials within view (such as a number line) will not provide visual aids to the child.

☑ Prepare the child for the test by explaining that the questions will be easy at first but will get harder and harder. You don't expect them to know all the answers. Some of them are even challenging for older children to get right!

☑ Administer *all* items at each level of the test that you progress to in order to determine the child's maximum potential within that level.

☑ Read or recite the test questions exactly as written, without rephrasing them. If the child appears not to understand or asks you to repeat the question, you may do so as many times as needed using the language of the test.

☑ Provide frequent reinforcement throughout, for example by saying "good" after each response, without actually telling children whether their answers are right or wrong.

Assessing Problem-Solving Strategy Usage

Like children's answers to the problems themselves, the strategies children use to solve these problems can aid you in determining their level of understanding. The arithmetic items in Levels 1–3 of the test include a reminder to "Assess strategy used by asking, 'How did you figure that out?'" It is helpful to ask the child how they arrived at the answer if this information is not apparent in the child's behavior. This will give you valuable information on the strategies a child has available to solve number problems.

For example, a child might rely on the earliest strategy to be acquired developmentally, which is starting from 1 and counting up. Or she might use a more sophisticated strategy and count on from the largest addend. Or she may retrieve the answer from memory and say, "I already knew the answer. It was in my head."

Use of fingers to solve problems can also indicate that the child has not yet constructed a mental counting line inside her head and must rely instead on physical objects. This valuable information suggests that the understandings that underlie the mental counting line (such as automatic knowledge of the number sequence, knowing which numbers come before/after other numbers, and knowing which numbers are larger/smaller than other numbers) must be carefully nourished and/or taught.

Recording Strategy Usage

The Number Knowledge Test Record includes a "Strategy" column to record the problem-solving strategy used by the child when answering the arithmetic items in Levels 1–3; those items are preceded by an asterisk (*) on the Test Record. To help you quickly record the strategies used on these selected items, the most common strategies (particularly at Level 1) are included on the Test Record in abbreviated form, as follows:

CU = Counts Up from 1. Circle "CU" if the child uses a counting up from 1 strategy for solving addition problems.

CO = Counts On. Circle "CO" if the child enters the counting sequence at the point of one of the addends and then counts on, one number at a time.

R = Retrieval. Circle "R" if the child says that she figured out the answer in her head, as in: "I just knew it," "My brain told me," or "I learned that."

Strategy Complexity

The three common strategies abbreviated on the Test Record vary in complexity. The CO strategy is more sophisticated than the CU strategy. When a child uses the CU strategy, she begins counting at 1 to count the first addend and continues counting up to add the second addend. Whereas when using the CO strategy, the child demonstrates a better understanding of the number sequence by beginning to count at the point in the sequence that is one number larger than the first addend. It often takes a child a year or more to move from one level of strategy use to another (e.g., from the CU strategy to the CO strategy or the from CO strategy to the R strategy for single-digit addition). Knowing which level of strategy use the child is comfortable using can provide valuable information for instructional planning.

Since children's strategies become more complex and varied at Levels 2 and 3 of the test, these common strategy abbreviations have been omitted on the Test Record at these higher levels to leave space for you to record more complex strategies used. If the child uses the CU, CO, or R strategy at these levels, use the abbreviations to record them. If a child uses a more complex strategy, create a shorthand way of recording it so that the child is not kept waiting.

For example, in the first item of Level 2 (What number comes 5 numbers after 49?), children who get the correct answer (54) often respond to the follow-up question (How did you figure that out?) in language similar to "50 and 5 is 55 but 50 is 1 more than 49 so I need to take 1 away to get 54." You can record this quickly with equations: $50 + 5 = 55; 55 - 1 = 54$. This strategy is more sophisticated than the CO strategy (which is also often used to solve this problem) because it demonstrates a good understanding of number composition and the base-ten concepts which underlie use of the *moving to 10 strategy* to solve double-digit addition problems.

While it is important to pay attention to the problem-solving strategies children use to solve problems, it is also important that the test-taking process proceeds smoothly and doesn't tire the child out too much. This is why only 3 items per level have been denoted to receive follow-up questions about strategy usage and most items have greyed out boxes in the "Strategy" column of the Test Record. If you find that even these questions seem to be burdensome for a child, you should use your discretion in omitting some or all of them. The test questions themselves are more important, and proceeding through the test until the child reaches a natural stopping point (indicated by directions on the test) is the ultimate goal.

Scoring the Test

If you begin testing at a level other than Preliminary/Level 0, give the child the maximum total of 5 points for the Level 0 items. Note that the maximum total points possible for each level is provided in the Test Record "Total" column. A reproducible Number Knowledge Test Record is included on page 32 of this book.

For all two-part items, both (a) and (b) must be answered correctly to earn a point. The rationale for this is that children can easily pick one of the two "forced-choice" response alternatives and obtain the correct answer by guessing. If the child answers both questions correctly, we can assume with greater confidence that she understands the number magnitude concepts assessed with these items.

The Number Knowledge Test Record

Total Score		Raw Test Score	
		Devel. Age Score	
Name	D.O.B.	Grade Level	
Date		NW Level	
School			
Teacher			

Preliminary (count 1–10)

Level 0 (3 or more correct, go to next level) — Strategy — Score

1. Count (3)			
2a. More: 5 vs. 2	**2b.** More: 3 vs. 7		
3a. Less: 2 vs. 6	**3b.** Less: 8 vs. 3		
4. Count B (4)			
5. Count A (8)		Total	/5

Level 1 (5 or more correct, go to next level) — Strategy — Score

* **1.** $4 + 3 =$		CU CO R	
2. $7 + 1 =$			
3. $7 + 2 =$			
4a. Bigger: 5 or 4	**4b.** Bigger: 7 or 9		
5a. Smaller: 8 or 6	**5b.** Smaller: 5 or 7		
6a. Closer to 5: 6 or 2	**6b.** Closer to 7: 4 or 9		
* **7.** $2 + 4 =$		CU CO R	
* **8.** $8 − 6 =$		CU CO R	
9a. First: 8 5 2 6	**9b.** Last: 8 5 2 6	Total	/9

Level 2 (5 or more correct, go to next level) — Strategy — Score

* **1.** $49 + 5 =$			
* **2.** $60 − 4 =$			
3a. Bigger: 69 or 71	**3b.** Bigger: 32 or 28		
4a. Smaller: 27 or 32	**4b.** Smaller: 51 or 39		
5a. Closer to 21: 25 or 18	**5b.** Closer to 28: 31 or 24		
6. How many #'s between 2 and 6			
7. How many #'s between 7 and 9			
* **8.** $12 + 54 =$		Add CO	Total
9. $47 − 21 =$			/9

Level 3 — Strategy — Score

* **1.** $99 + 10 =$			
2. $99 + 9 =$			
3a. Bigger: $9 − 6$ or $8 − 3$	**3b.** Bigger: $6 − 2$ or $8 − 5$		
4a. Smaller: $99 − 92$ or $25 − 11$	**4b.** Smaller: $48 − 36$ or $84 − 73$		
* **5.** $13 + 39 =$			
* **6.** $36 − 18 =$		Total	
7. $301 − 7 =$			/7

* Denotes arithmetic item

32 Number Knowledge Test Record

Using the Test Record

☑ Each item on the test is worth 1 point. Earning partial credit is not allowable. In multi-part items, both the "a" and "b" questions must be correct in order to earn a point.

☑ In the "Score" column, record either a 1 for a correct answer or a 0 for an incorrect answer for each item.

☑ When recording an item's score, mark it in such a way that failures are not apparent to the child.

☑ Once all items in a level have been scored, tally the points in the "Total" column to determine whether to discontinue testing or move on to the next level. The amount of points needed to proceed to the next level is indicated next to the level number on the Test Record.

☑ If the child is not able to progress to the next level(s) of the test, the child receives 0 points for those level(s).

☑ Add the points earned for all levels of the test and record them in the "Raw Test Score" box.

☑ If possible, enter the child's answer (whether correct or incorrect) on the Test Record in the space after each item's question. This data can provide valuable information on the level of understanding when you have a chance to analyze it.

☑ Arithmetic items are preceded by an asterisk (*). For these items, record the strategy the child used to solve the problem under the "Strategy" column.

☑ Using the Raw Test Score, fill in the "Developmental Age Score," "Grade Level," and "Number Worlds Level" based on the Developmental Conversion Chart:

Raw Test Score	Developmental Age Score (Chronological Age Equivalents)	Grade Level Equivalents	Number Worlds Level
1–6	3–4 years	Preschool	Level A
7–8	4–5 years	PreK–K	Level B
9–14	5–6 years	K–1	Level C
15–19	6–7 years	1–2	Level D
20–25	7–8 years	2–3	Level E
26–28	8–9 years	3–4	Level F
29–30	9–10 years	4–5	Level G

☑ The Developmental Age score provides a measure of the child's pre-instruction level of number sense or math competence.

☑ The Number Worlds Level indicates the level of Placement Test the child should be given to further refine and affirm their placement within the **Number Worlds** program.

Using the Test Results

The results of The Number Knowledge Test can be used for four main purposes:

To determine if a child is functioning at, above, or below age level in number knowledge:

If the child's Developmental Age Score is consistent with her chronological age, you can infer that her number knowledge is *average* for her age and consistent with about 64% of her age-level peers in North America. If it is one or two levels higher, which is the case for about 18% of American children, you can infer that her number knowledge is *above average*. Similarly, if it is one or two levels lower, which is the case for another 18% of American children, you can infer that her number knowledge is *below average* and remedial instruction is needed.

As mentioned, many children from low-income communities, lacking the learning opportunities available to their more advantaged peers, have been found to score in the *below average* range on this test. These children typically benefit hugely from instruction with the **Number Worlds** program, which was designed to fill the knowledge gaps demonstrated on this test.

To determine which number concepts the child has mastered, which she is struggling with, and which she still needs to learn:

The Number Knowledge Test assesses four classes of concepts at almost every level of the test: (1) number sequence concepts, including base-ten understandings; (2) addition concepts; (3) subtraction concepts; and (4) math problem-solving strategies. By examining the completed Test Record and evaluating the particular items the child passed and failed at each level of the test, you can gain a good picture of the child's knowledge strengths and weaknesses and identify concepts that will need extra attention in your instructional programming.

Many children demonstrate weaknesses in the areas of number sequence and subtraction understandings, most likely because these concepts have not been adequately addressed in the child's home, community, or school, and the **Number Worlds** program contains several lessons at every level to teach these important concepts.

If you have asked the strategy-usage question on the indicated arithmetic items while administering the test and recorded the child's answer, you will also have valuable information available to determine the sophistication of the child's problem-solving strategies and the strategies that need to be taught and/or fostered.

To assess progress over the instructional period or academic year:

Administer The Number Knowledge Test at the beginning and end of the instructional period or academic year, making sure that the interval between test administrations is at least 4 months. By comparing the Developmental Age Score a child achieved before instruction began to the score received at the end of the instructional period, you can obtain a measure of the developmental growth of the child over the instructional period. You can also compare the scores of individual levels or items to identify concepts the child mastered over the instructional period and those she has yet to master.

If a child moved up one level on the test, you can infer that they are now performing at a level that is one year above the level they started at. Similarly, if they move up two levels in their Developmental Age Score, you can conclude that they are now performing at a level that is two years above the level at which they began. Identifying areas of growth can give you a well-deserved sense of accomplishment and a topic of welcomed feedback for the child's parents. If growth is not as great as you expected or hoped for in some areas, you can use these results to modify or adapt your instructional planning for the next school year.

By comparing the percentage of children in your class who passed each individual item on the test before and after the instructional period, you can obtain a good index of the particular concepts your class mastered during the instructional period as well as the particular concepts they are still struggling with and have yet to master. This information can provide valuable guidance for all teachers who will be working with your class in the coming months and/or year.

To determine which *Number Worlds* level Placement Test to start with and which level of the *Number Worlds* program to use to begin instruction:

The **Number Worlds** Placement Tests can help you refine your assessment of the child's number competence and pinpoint which level of the **Number Worlds** program a child should begin working within. To determine which Placement Test to start with, use the Developmental Conversion Chart and find the **Number Worlds** level that is associated with the Raw Score the child achieved. More information about using and administering Placement Tests is provided on pages 34–37 of this book.

FAQs about Test Items

Question about test administration: Why do we have to ask *all* items at any level of the test the child progresses to if they fail the first few items at this level? Won't this be frustrating to the child?

Answer: Although each level of the test is more difficult than the previous level, the test questions within each level are not sequenced in order of difficulty. They are sequenced to maximize comprehension of the test question. Thus, a child could easily fail the first few questions at a level and succeed on subsequent questions at this level. To ensure that you give the child every opportunity to demonstrate her number knowledge, it is imperative that you ask all questions at any level you progress to.

Question about the Visual Array in item 6a at Level 1: Why do we show a visual array? Won't this lead children to try to measure the physical distance between numbers on the page?

Answer: The visual array was included to provide a counterweight to some children's tendency to choose the number that is, auditorily, closer to the number said in the oral test question. You are correct in assuming that a few children will try to physically measure the spatial distance between numbers when the visual array is used. However, they won't even think of doing this if they understand that numbers themselves have magnitude, and it is this understanding that the test item was designed to measure.

Question about the Visual Array in item 9a at Level 1: Isn't it unfair to show children written numbers that are not sequenced in order of magnitude and ask them to point to the number they say first when counting? Aren't you deliberately distracting them?

Answer: Yes, the distraction is deliberate and is consistent with many items on developmental tests. It is designed to assess the solidity of a child's understanding of the counting sequence. If their knowledge is solid, they won't be distracted. The fact that this item has remained on the test after substantial research to ensure that the test is valid suggests that most children in the 6-year-old age range are able to pass this item in its current form.

Question about item 6 at Level 2: Why should we accept either 3 or 4 as a correct answer?

Answer: If the child understands the number sequence, she will be able to figure out that there are 3 numbers between 2 and 6: namely, the numbers 3, 4, and 5. However, if the child uses subtraction to find the answer, she will take 2 away from 6 and come up with an answer of 4. Although the question is worded to test number sequence understandings, we don't want to penalize a child for demonstrating a good understanding of subtraction so we accept either answer as correct. By noting which strategy the child uses to solve this problem, you can determine whether number sequence understandings need to be taught.

Question about item 3a at Level 3: The wording of this question seems very difficult for children, especially ESL children who may not know what the word "difference" means. Can we reword this question to make it easier to understand?

Answer: No, you cannot reword this question but you may repeat it as often as requested by the child. If the child is functioning at the 10-year-old level mathematically, she will understand what this word means. If she doesn't, this failure suggests that she is not functioning at age level and needs remedial instruction in subtraction understandings.

This question also imposes demands on memory, and this adds to the level of difficulty. However, most children who are functioning at the 10-year-old level are able to handle this memory challenge. Failure to do so provides an indication that the child's mental capacity (and possibly English language skills) for mathematical problem-solving is below average and remedial instruction to foster the development of these skills is needed.

The Number Knowledge Test

Preliminary

Let's see if you can count from 1 to 10. Go ahead.

Level 0 (4-year-old level):

Go to Level 1 if 3 or more are correct

1. Can you count these Counters and tell me how many there are?
(Place 3 Counters in a row in front of the child.)

2a. (Show stacks of Counters, 5 vs. 2, same color.)
Which pile has more?

2b. (Show stacks of Counters, 3 vs. 7, same color.)
Which pile has more?

3a. This time, I'm going to ask you which pile has less.
(Show stacks of Counters, 2 vs. 6, same color.)
Which pile has less?

3b. (Show stacks of Counters, 8 vs. 3, same color.)
Which pile has less?

4. I'm going to show you some Counters.
(Show a line of 3 Counters of one color [A] and 4 Counters of a different color [B] in a row, as follows: A B A B A B B.)
Count just the (color B) Counters and tell me how many there are.

5. (Clear all the Counters from the previous question. Show a mixed array—not a row—of 8 Counters of one color [A] and 7 Counters of a different color [B].)
Here are some more Counters. Count just the (color A) Counters and tell me how many there are.

The Number Knowledge Test

Level 1 (6-year-old level):

Go to Level 2 if 5 or more are correct

1. If you had 4 chocolates and someone gave you 3 more, how many chocolates would you have all together?
(Assess strategy used by asking, "How did you figure that out?")

2. What number comes right after 7?

3. What number comes two numbers after 7?

4a. Which is bigger: 5 or 4?

4b. Which is bigger: 7 or 9?

5a. This time, I'm going to ask you about smaller numbers. Which is smaller: 8 or 6?

5b. Which is smaller: 5 or 7?

6a. Which number is closer to 5: 6 or 2?
(Show Visual Array 1 after asking the question)

6b. Which number is closer to 7: 4 or 9?
(Show Visual Array 2 after asking the question.)

7. How much is 2 + 4?
(Children can use fingers for counting. Assess strategy used by asking, "How did you figure that out?")

8. How much is 8 take away 6?
(Children can use fingers for counting. Assess strategy used by asking, "How did you figure that out?")

9a. (Show Visual Array 3 of the numerals 8 5 2 6, and ask the child to point to and name each numeral.)
When you are counting, which of these numbers do you say first?

9b. When you are counting, which of these numbers do you say last?

The Number Knowledge Test

Level 2 (8-year-old level):

Go to Level 3 if 5 or more are correct

1.   What number comes 5 numbers after 49?
(Assess strategy used by asking, "How did you figure that out?")

2.   What number comes 4 numbers before 60?
(Assess strategy used by asking, "How did you figure that out?")

3a. Which is bigger: 69 or 71?

3b. Which is bigger: 32 or 28?

4a. This time I'm going to ask you about smaller numbers. Which is smaller: 27 or 32?

4b. Which is smaller: 51 or 39?

5a. Which number is closer to 21: 25 or 18?
(Show Visual Array 4 after asking the question.)

5b. Which number is closer to 28: 31 or 24?
(Show Visual Array 5 after asking the question.)

6.   How many numbers are there between 2 and 6?
(Accept either 3 or 4.)

7.   How many numbers are there between 7 and 9?
(Accept either 1 or 2.)

8.   How much is 12 + 54?
(Show Visual Array 6. Assess strategy used by asking, "How did you figure that out?")

9.   How much is 47 take away 21?
(Show Visual Array 7.)

The Number Knowledge Test

Level 3 (10-year-old level):

1. What number comes 10 numbers after 99?
(Assess strategy used by asking, "How did you figure that out?")

2. What number comes 9 numbers after 99?

3a. Which difference is bigger, the difference between 9 and 6 or the difference between 8 and 3?

3b. Which difference is bigger, the difference between 6 and 2 or the difference between 8 and 5?

4a. Which difference is smaller, the difference between 99 and 92 or the difference between 25 and 11?

4b. Which difference is smaller, the difference between 48 and 36 or the difference between 84 and 73

5. How much is 13 + 39?
(Show Visual Array 8. Assess strategy used by asking, "How did you figure that out?")

6. How much is 36 − 18?
(Show Visual Array 9. Assess strategy used by asking, "How did you figure that out?")

7. How much is 301 take away 7?

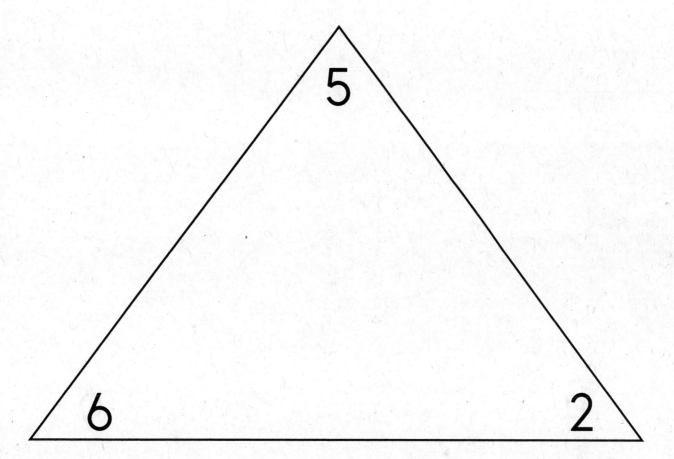

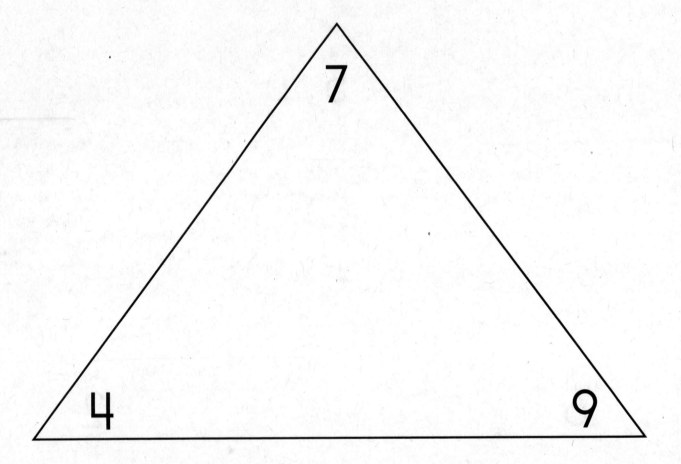

8   5   2   6

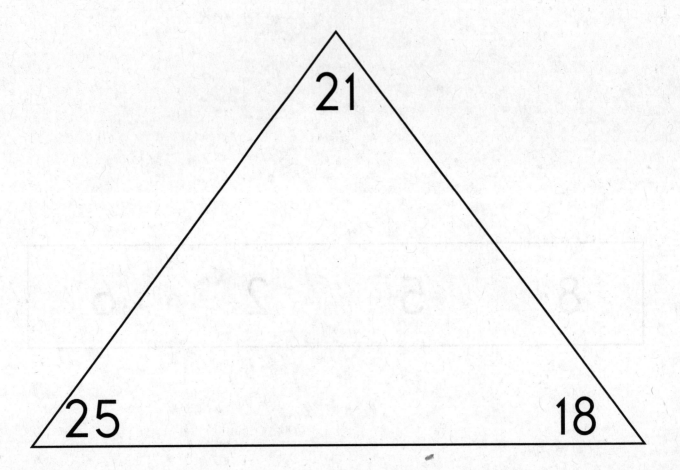

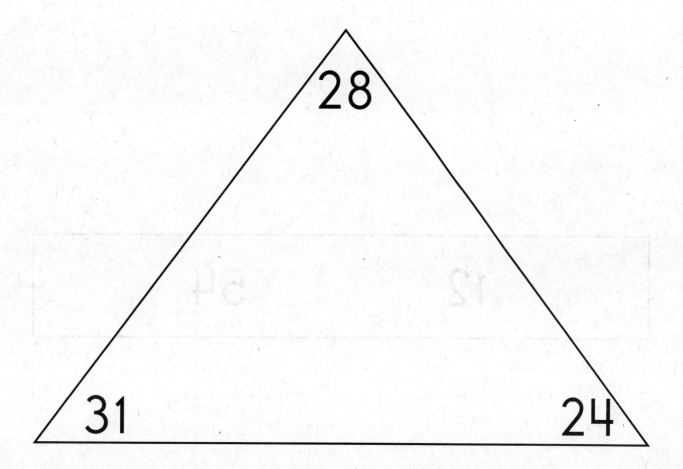

12	54

47   21

13   39

36   18

The Number Knowledge Test Record

Total Score		Raw Test Score	
		Devel. Age Score	
Name	**D.O.B.**	**Grade Level**	
Date		**NW Level**	
School			
Teacher			

Preliminary (count 1–10)

Level 0 (3 or more correct, go to next level)

				Strategy	Score
1.	Count (3)				
2a.	More: 5 vs. 2	**2b.**	More: 3 vs. 7		
3a.	Less: 2 vs. 6	**3b.**	Less: 8 vs. 3		
4.	Count B (4)				
5.	Count A (8)				

Total /5

Level 1 (5 or more correct, go to next level)

					Strategy	Score
*	**1.**	$4 + 3 =$			CU CO R	
	2.	$7 + 1 =$				
	3.	$7 + 2 =$				
	4a.	Bigger: 5 or 4	**4b.**	Bigger: 7 or 9		
	5a.	Smaller: 8 or 6	**5b.**	Smaller: 5 or 7		
	6a.	Closer to 5: 6 or 2	**6b.**	Closer to 7: 4 or 9		
*	**7.**	$2 + 4 =$			CU CO R	
*	**8.**	$8 - 6 =$			CU CO R	
	9a.	First: 8 5 2 6	**9b.**	Last: 8 5 2 6		

Total /9

Level 2 (5 or more correct, go to next level)

					Strategy	Score
*	**1.**	$49 + 5 =$				
*	**2.**	$60 - 4 =$				
	3a.	Bigger: 69 or 71	**3b.**	Bigger: 32 or 28		
	4a.	Smaller: 27 or 32	**4b.**	Smaller: 51 or 39		
	5a.	Closer to 21: 25 or 18	**5b.**	Closer to 28: 31 or 24		
	6.	How many #'s between 2 and 6				
	7.	How many #'s between 7 and 9				
*	**8.**	$12 + 54 =$			CO	
	9.	$47 - 21 =$				

Total /9

Level 3

					Strategy	Score
*	**1.**	$99 + 10 =$				
	2.	$99 + 9 =$				
	3a.	Bigger: $9 - 6$ or $8 - 3$	**3b.**	Bigger: $6 - 2$ or $8 - 5$		
	4a.	Smaller: $99 - 92$ or $25 - 11$	**4b.**	Smaller: $48 - 36$ or $84 - 73$		
*	**5.**	$13 + 39 =$				
*	**6.**	$36 - 18 =$				
	7.	$301 - 7 =$				

Total /7

* Denotes arithmetic item

Using the *Number Worlds* Placement Tests

The **Number Worlds** Placement Tests are used to determine in which **Number Worlds** level students should begin their instruction. There is a Placement Test per level in levels B-I (with level A being the level of instruction for students who do not pass Placement Test B and level J being the level of instruction for students who excel on Placement Test I). Every Placement Test assesses students' preexisting knowledge of the Common Core State Standards associated with that level. The items in the tests are arranged from easiest to most difficult. A student's score on a Placement Test will indicate whether or not they need to continue testing at a different level so that the appropriate level of instruction within the program can be determined.

If you have given the Number Knowledge Test to the student, it is recommended that you administer the Placement Test that corresponds to their Raw Test Score on the following Developmental Conversion Chart. If you have not given the student the Number Knowledge Test, locate the student's current grade in the Grade Level Equivalents column, and administer the corresponding level Placement Test. The Grade Level Equivalents cover a range of two grade levels (e.g. K-1). Begin testing with the lower Placement Test. For example, if your student is in the 3rd grade, begin testing with Placement Test E. This will help ensure that children achieve success on the beginning items of this test and provide a baseline measure of the child's math competence.

Developmental Conversion Chart

Raw Test Score	Developmental Age Score (Chronological Age Equivalents)	Grade Level Equivalents	Number Worlds Level
1–6	3–4 years	Preschool	Level A
7–8	4–5 years	PreK-K	Level B
9–14	5–6 years	K–1	Level C
15–19	6–7 years	1–2	Level D
20–25	7–8 years	2–3	Level E
26–28	8–9 years	3–4	Level F
29–30	9–10 years	4–5	Level G
N/A	10–11 years	5–6	Level H
N/A	11–12 years	6–7	Level I
N/A	12–13 years	7–8	Level J

Critical Range Testing

The *Number Worlds* Placement Tests are collectively and individually a critical range test. A critical range test consists of a series of items arranged from easiest to most difficult. A student completes a portion (or range) of items in the test which approximates their abilities. If a student makes several consecutive correct responses, it can be assumed with confidence that the student would also respond correctly to the preceding items, which are easier. Therefore, those items need not be administered. Similarly, when a student makes several incorrect responses, it can be assumed that the student would also not have the capability to correctly answer the subsequent items, which are more difficult.

This progression of difficulty also holds true across levels, so many students will need to be given more than one level's Placement Test in order to assess their current level of understanding. Using a critical range test for assessment is an ideal approach for placement within the program since it is neither practical for the teacher nor desirable for the student to provide a full range of items to all examinees. Only items estimated to be within the student's current probable range of math understanding are administered. This approach saves time in the testing process and prevents the student from becoming overly frustrated.

The items used in the Placement Tests are representative of the content and standards covered within that level. Effort has been taken to use language that should be familiar to most students instead of *Number Worlds* specific terminology. The Developmental Conversion Chart can be used as a guide for determining the appropriate Placement Test with which to begin testing, but generally speaking, you want to underestimate a student's ability because it is preferable that the student has the opportunity to succeed in the beginning of the test-taking process. If your previous experience with a student suggests to you that they will perform below their current grade level, you should begin testing at the level you deem appropriate based on the information contained in the Developmental Conversion Chart.

Placement Tests may also be given at the end of the year (or once the *Number Worlds* curriculum for a given level is completed) and compared to the students' initial scores in order to measure the students' overall progress in their understanding of math concepts and Common Core State Standards.

Placement Tests in Levels A–C

Number Worlds levels A–C are targeted for use by students in grades Pre-Kindergarten through Grade 1. Placement Tests can be used to identify where students should begin within the *Number Worlds* curriculum.

Placement Tests for levels B and C are designed to be administered individually to each student by a teacher, classroom aid, or parent helper. The tests at these levels consist of teacher's instructions on the left-hand page and reproducible test masters on the right-hand page.

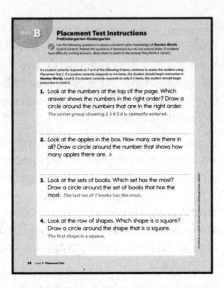

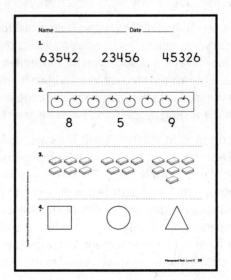

Placement Test, Level B

Placement Tests B and C contain 8 questions apiece which can be used to assess a student's prior knowledge of the corresponding *Number Worlds* content and Common Core State Standards for those levels. You may repeat questions to the students if necessary, but do not reword them.

- If a student correctly responds to 7 or 8 of the items, continue to assess the student using the next Placement Test up.

- If a student correctly responds to 4–6 items, the student should begin instruction within that *Number Worlds* level.

- If a student correctly responds to only 0–3 items in Placement Test C, continue to assess the student using Placement Test B.

- If a student correctly responds to only 0–3 items in Placement Test B, the student should begin instruction within *Number Worlds* level A.

To gain an even more thorough understanding of a student's conceptual knowledge of number (number sense) at these levels, use the Number Knowledge Test included in this book (pp. 19–32). The Number Knowledge Test was designed to assess central conceptual knowledge typically acquired by children around the ages of 4, 6, 8, and 10 years.

Placement Tests in Levels D–J

Number Worlds levels D–J are targeted for use by students in Grades 2-8. Placement of students at these levels within the **Number Worlds** curriculum should begin by evaluating their success on the reproducible Placement Tests on pages 46–63 of this book.

At these levels, students should attempt to take the test independently. If a student is struggling with reading comprehension of the items, the Placement Tests may be administered by a teacher, classroom aide, or parent helper. In order to best evaluate the effectiveness of the **Number Worlds** program and prepare the student for future testing, students should complete the test on their own.

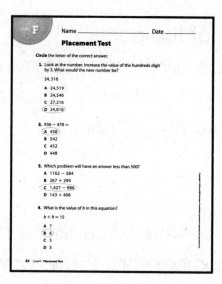
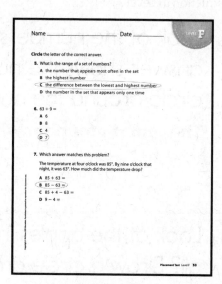

Placement Test, Level F

Placement Tests D–I contain 10 questions apiece which can be used to assess a student's prior knowledge of the corresponding **Number Worlds** content and Common Core State Standards for those levels. Ideally, a student should get most of the items correct in the level in which they are placed.

- If a student correctly responds to 8–10 of the items, continue to assess the student using the next Placement Test up.

- If a student correctly responds to 8–10 of the items in Placement Test I, the student should begin instruction within **Number Worlds** level J.

- If a student correctly responds to 4–7 items, the student should begin instruction within that **Number Worlds** level.

- If a student correctly responds to only 0–3 items, continue to assess the student using the next Placement Test down.

Placement Test Instructions
PreKindergarten-Kindergarten

✓ Use the following questions to assess a student's prior knowledge of *Number Worlds* Level B content. Repeat the questions if necessary but do not reword them. If students have difficulty circling answers, allow them to point to the answer they think is correct.

If a student correctly responds to 7 or 8 of the following 8 items, continue to assess the student using Placement Test C. If a student correctly responds to 4-6 items, the student should begin instruction in *Number Worlds*, Level B. If a student correctly responds to only 0-3 items, the student should begin instruction in Level A.

1. Look at the numbers at the top of the page. Which answer shows the numbers in the right order? Draw a circle around the numbers that are in the right order.

The center group showing 2 3 4 5 6 is correctly ordered.

2. Look at the apples in the box. How many are there in all? Draw a circle around the number that shows how many apples there are. **8**

3. Look at the sets of books. Which set has the most? Draw a circle around the set of books that has the most. **The last set of 7 books has the most.**

4. Look at the row of shapes. Which shape is a square? Draw a circle around the shape that is a square.

The first shape is a square.

Name _____ Date _____

1.

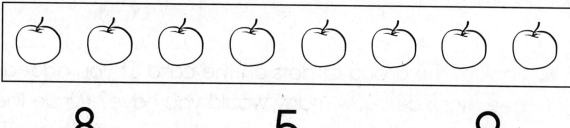

63542   23456   45326

2.

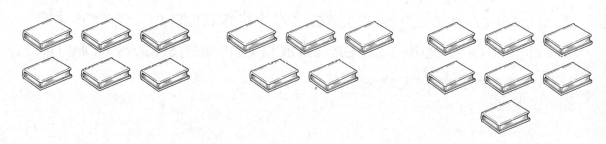

8   5   9

3.

 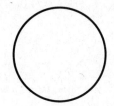 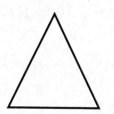

4.

Placement Test Instructions
PreKindergarten-Kindergarten

Use the following questions to assess a student's prior knowledge of **Number Worlds** Level B content. Repeat the questions if necessary but do not reword them. If students have difficulty circling answers, allow them to point to the answer they think is correct.

5. Look at the numbers at the top of the page. Which number is the largest? Draw a circle around the number that is the largest. 9

6. Look at the group of dots on the card. If you added one more dot, how many would you have? Circle the answer that shows how many dots you would have if you added one more to the dots on the card? 6

7. Look at the crayons. Raj had this many crayons. He gave one to his sister. Which answer shows how many crayons Raj had left? 9

8. What number will the star be on if you move forward five spaces? 9

Name _____ Date _____

5.

3   9   7

- -

6.

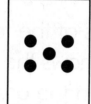

4   8   6

- -

7.

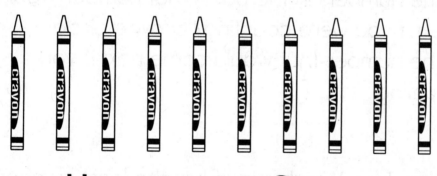

4   9   7

- -

8.

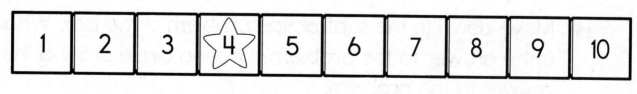

5   7   9

LEVEL C

Placement Test Instructions
Kindergarten-Grade 1

Use the following questions to assess a student's prior knowledge of **Number Worlds** Level C content. Repeat the questions if necessary but do not reword them.

If a student correctly responds to 7 or 8 of the following 8 items, continue to assess the student using Placement Test D. If a student correctly responds to 4-6 items, the student should begin instruction in **Number Worlds**, Level C. If a student correctly responds to only 0-3 items, continue to assess the student using Placement Test B.

1. Look at the numbers at the top of the page. Which group of numbers shows the correct counting order? Draw a circle around the group of numbers that shows the correct counting order.

The last group showing 16, 17, 18, 19 is correctly ordered.

2. Look at the numbers in the box. What number would come next if you were counting? Draw a circle around the number that would come next if you were counting. **31**

3. Look at the dogs. Which group has the least number of dogs? Draw a circle around the group that has the least number of dogs. **The first group of 5 dogs has the least.**

4. Move down to the subtraction problem in the box. What is the answer to the problem? Draw a circle around the answer to the problem. **7**

Name _____ Date _____

1.

15, 13, 14, 17   14, 12, 16, 15   16, 17, 18, 19

2.

| 28, 29, 30, ___ |

31   27   33

3.

4.

| 16 − 9 = ___ |

7   8   9

Placement Test Instructions
Kindergarten-Grade 1

✓ Use the following questions to assess a student's prior knowledge of *Number Worlds* Level C content. Repeat the questions if necessary but do not reword them.

5. Listen carefully and look at the next row of numbers. Ann had five stickers. She put four stickers in a book and one on a letter. How many did she have left? Draw a circle around the answer that shows how many stickers she had left. 0

6 Look at the problem in the box. What is the answer to the problem? Draw a circle around the number that is the answer to the problem. 6

7 Look at the problem in the next box. What is the answer to the problem? Draw a circle around the number that is the answer to the problem. 19

8 How many tens are in the number 52? 5

Name _____ Date _____

5.

5   1   0

- -

6.

$$2 + 1 + 3 = \underline{\quad}$$

4   6   5

- -

7.

$$13 + 6 = \underline{\quad}$$

19   14   18

- -

8.

2   5   7

Name _____ Date _____

Placement Test

Circle the letter of the correct answer.

1. Which answer is the same as 8 + 6 + 9?

    A 8 tens and 6 ones

    B 6 tens and 9 ones

    C 2 tens and 3 ones

    D 3 tens and 8 ones

2. Look at the squares. How many groups of tens and ones are there?

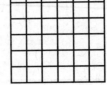

    A 6 tens and 6 ones

    B 2 tens and 4 ones

    C 3 tens and 6 ones

    D 5 tens and 4 ones

3. What shape is missing from this pattern?

      _____

    A

    B

    C

    D

Name _____ Date _____

Circle the letter of the correct answer.

4. What number goes in the box to make the problem correct?

$$\Box + 14 = 19$$

A 6   **B** 8

C 7   **D** 5

5. Look at the number line. If the rule continues, where will the arrow land on the next hop?

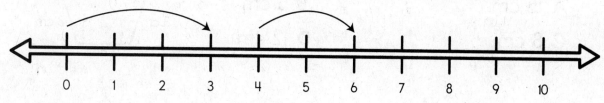

A 7   **B** 9

C 10   **D** 13

6. Which answer is an equation?

A $9 = 3 = 6$

B $2 + 4 + 7$

C $7 - 2 + 7$

D $8 + 3 = 11$

7. Pat saw 5 fish and 3 turtles in a pond. Then 2 frogs hopped into the pond. How many things in all were in the pond?

A 10   **B** 8

C 12   **D** 7

Name _____ Date _____

Circle the letter of the correct answer.

8. Which equation matches this problem?

There were 9 birds in a tree. All of them flew away.
How many birds were left in the tree?

A $9 + 1 = 10$ **B** $9 - 9 = 0$

C $4 + 5 = 9$ **D** $9 + 0 = 9$

9. What is the perimeter of this shape?

A 15 cm   **B** 5 cm

C 8 cm   **D** 12 cm

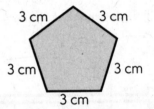

10. Look at the graph. How many miles did Mom run?

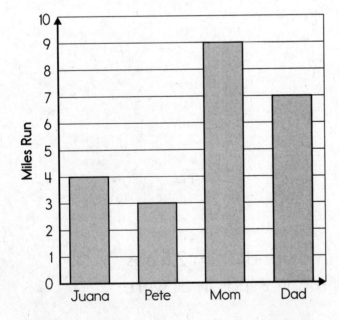

A 7   **B** 9

C 6   **D** 10

Name _____ Date _____

Placement Test

Circle the letter of the correct answer.

1. Which answer is the same as 6 tens and 27 ones?

   A 33

   B 87

   C 62

   D 82

2. How many dimes are equal to 40 pennies?

   A 10

   B 400

   C 14

   D 4

3. Which answer is correct?

   A $5 = 13$

   B $4 < 9$

   C $8 < 3$

   D $2 > 12$

4. $9 + 7 =$ _____

   A $9 - 7$

   B $9 + (7 + 9)$

   C $7 + (6 + 3)$

   D $7 - (9 + 3)$

Name _____ Date _____

Circle the letter of the correct answer.

5. In which problem do you have to regroup?

    A 61 − 19

    B 54 − 42

    C 38 − 25

    D 76 − 30

6. Look at the pictograph. How many cars crossed the bridge on Wednesday?

Number of Cars Crossing Bridge

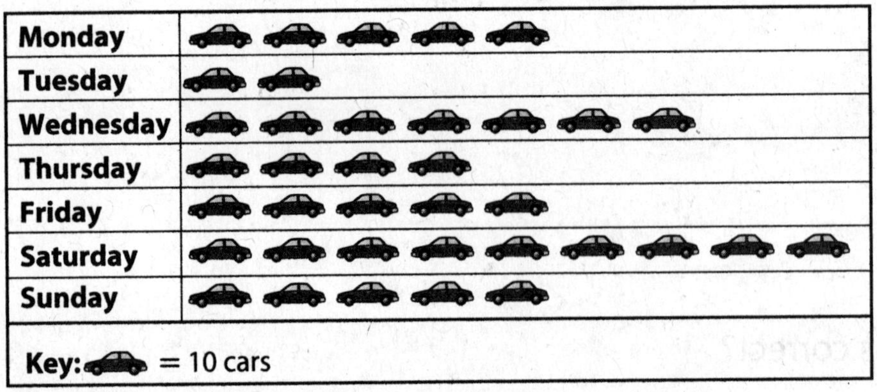

Monday	🚗 🚗 🚗 🚗 🚗
Tuesday	🚗 🚗
Wednesday	🚗 🚗 🚗 🚗 🚗 🚗 🚗
Thursday	🚗 🚗 🚗 🚗
Friday	🚗 🚗 🚗 🚗 🚗
Saturday	🚗 🚗 🚗 🚗 🚗 🚗 🚗 🚗 🚗
Sunday	🚗 🚗 🚗 🚗 🚗

Key: 🚗 = 10 cars

    A 17

    B 24

    C 53

    D 70

7. What is the value of the digit 3 in the number 539?

    A 13

    B 30

    C 39

    D 300

Name _____ Date _____

Circle the letter of the correct answer.

8. What number is missing in this function table?

IN	OUT
3	8
5	10
7	
15	20
23	28

A 33   **B** 14

C 29   **D** 12

9. A point on this graph is missing. The *x*-value of the point is 4. What is the *y*-value?

A 3   **B** 20

C 14   **D** 7

10. Which equation is correct?

A $5 + 9 = 14 - 5$

B $5 + 5 = 9 + 9$

C $5 + 9 = 9 + 5$

D $9 + 5 = 14 - 9$

Name _____ Date _____

Placement Test

Circle the letter of the correct answer.

1. Look at the number. Increase the value of the hundreds digit by 3. What would the new number be?

   24, 516

   A 24,519

   B 24,546

   C 27,216

   D 24,816

2. $936 - 478 =$

   A 458

   B 542

   C 452

   D 448

3. Which problem will have an answer less than 500?

   A $1162 - 584$

   B $267 + 294$

   C $1,427 - 986$

   D $143 + 406$

4. What is the value of b in this equation?

   $b + 9 = 15$

   A 7

   B 6

   C 3

   D 5

Name _____ Date _____

Circle the letter of the correct answer.

5. What is the range of a set of numbers?

   A the number that appears most often in the set

   B the highest number

   C the difference between the lowest and highest number

   D the number in the set that appears only one time

6. $63 \div 9 =$

   A 6

   B 8

   C 4

   D 7

7. Which answer matches this problem?

The temperature at four o'clock was 85°. By nine o'clock that night, it was 63°. How much did the temperature drop?

   A $85 + 63 =$

   B $85 - 63 =$

   C $85 + 4 - 63 =$

   D $9 - 4 =$

Name _____ Date _____

Circle the letter of the correct answer.

8. Look at the graph. In which game did Lee score more points than Chris?

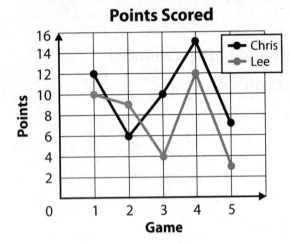

Points Scored

A 1

B 2

C 3

D 4

9. A driver delivered 214 soccer balls to a store. The next day, the driver brought 27 footballs and 39 basketballs to the same store. If 25 balls can fit on a shelf, how many shelves will be needed for all the balls?

A 13

B 15

C 12

D 19

10. What is the area of a room that is 8 feet wide and 12 feet long?

A 96 square feet

B 20 square feet

C 128 square feet

D 82 square feet

Name _____ Date _____

Placement Test

Circle the letter of the correct answer.

1. Which figure has the smallest surface area?

A

B

C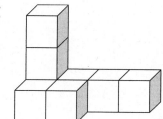

D

2. $\frac{3}{4} \times \frac{2}{3} =$

   A $\frac{3}{4}$

   B $\frac{2}{3}$

   C $\frac{1}{4}$

   D $\frac{1}{2}$

Name _____ Date _____

Circle the letter of the correct answer.

3. $\frac{1}{4} + \frac{1}{3} + \frac{1}{12} =$

   A $\frac{1}{4}$

   B $\frac{5}{12}$

   C $\frac{2}{3}$

   D $\frac{2}{7}$

4. What is the median of the following set of numbers?

   11 2 5 6 7 14 4

   A 3

   B 5

   C 6

   D 7

5. A board is 8 feet long. There are 12 inches in a foot. A worker cut the board into 6 pieces. How long was each piece of wood?

   A 16 inches

   B 14 inches

   C 4 inches

   D 48 inches

6. $6 \times (2 + 5) =$

   A 42

   B 13

   C 17

   D 67

Circle the letter of the correct answer.

7. Which answer is equivalent to $\frac{1}{4}$?

   A $\frac{1}{8}$

   B $\frac{4}{14}$

   C $\frac{14}{28}$

   D $\frac{8}{32}$

8. Which answer completes the equation?

   $33 - 8 = \underline{\hphantom{mm}} + 9$

   A 25

   B 17

   C 16

   D 34

9. There are three times more cows in a field than sheep. If there are 40 animals all together, how many of them are sheep?

   A 12

   B 10

   C 4

   D 20

10. Which answer is equal to 2^3?

   A 2×3

   B 3^2

   C $2 \times 2 \times 2$

   D $1 \times 2 \times 3$

Name _____ Date _____

Placement Test

Circle the letter of the correct answer.

1. In which answer are the numbers ordered from least to greatest?

 A −2, −3, −5, 9, 13

 B −12, −4, 2, 7, 14

 C 2, −2, 3, −3, 5, −5

 D 18, 12, 0, −6, −1

2. The table below shows the pets owned by students in a class. What is the ratio of fish to cats?

PET	NUMBER
Fish	4
Birds	2
Dogs	9
Cats	7

 A 11   **B** 3

 C $\frac{4}{2}$ **D** $\frac{4}{7}$

3. What is the volume of this figure?

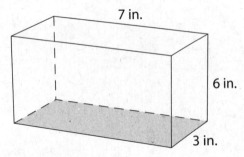

 A 126 in^3

 B 16 in^3

 C 367 in^3

 D 63 in^3

Name _____ Date _____

Circle the letter of the correct answer.

4. $358\overline{)14962}$

   A   44

   B   42 R 9

   C   41 R 284

   D   14 R 962

5. If you divided a circle into 3 equal angles, what would each angle measure?

   A   300 degrees

   B   120 degrees

   C   90 degrees

   D   63 degrees

6. What is 250% of 40?

   A   100

   B   254

   C   425

   D   10

7. $\frac{2}{5} \div \frac{1}{8} =$

   A $2\frac{5}{8}$

   B $\frac{3}{13}$

   C $\frac{5}{16}$

   D $3\frac{1}{5}$

Name _____ Date _____

Circle the letter of the correct answer.

8. What is the measure of the missing interior angle in this triangle?

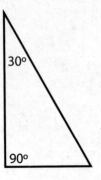

A   39 degrees

B   60 degrees

C   40 degrees

D   120 degrees

9. What are the coordinates of point *c*?

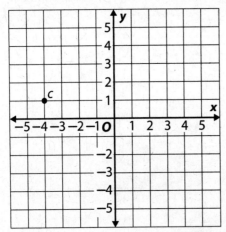

A   (2, −2)

B   (−4, 1)

C   (1, −3)

D   (−4, −4)

10. A box contains three balls of different colors. The colors are red, white, and blue. What is the probability of choosing the same color ball two times in a row?

A $\frac{2}{3}$

B $\frac{1}{9}$

C $\frac{1}{3}$

D $\frac{2}{27}$

Name _____ Date _____

Placement Test

Circle the letter of the correct answer.

1. What is the least common multiple of these numbers?

3, 4, and 9

A 12

B 36

C 18

D 49

2. Which numbers are correctly ordered from least to greatest?

A $-\frac{5}{3}, -1, 0.05, 2\frac{1}{3}, \frac{9}{2}$

B $0, 1, \frac{2}{3}, -\frac{4}{2}, -0.7$

C $-0.5, -\frac{2}{3}, 1, \frac{5}{4}, 0.07$

D $\frac{1}{4}, -0.4, 2, -\frac{3}{5}, 4$

3. $63 - (-19) = $ _____

A 44

B 56

C 82

D 73

Name _____ Date _____

Circle the letter of the correct answer.

4. $11\frac{1}{3} - 7\frac{3}{5} =$ _____

A $3\frac{11}{15}$

B $4\frac{1}{4}$

C 18

D $4\frac{1}{5}$

5. The regular price of a soccer ball is $28. It is on sale for 25% off. What is the sale price of the soccer ball?

A $21

B $3

C $7

D $25

6. $6.75 \div 0.25 =$ _____

A 6.5

B 4.5

C 13

D 27

7. Which of these proportions is not equal to the others?

A 1:3

B 4:5

C 15:45

D 6:18

Name _____ Date _____

Circle the letter of the correct answer.

8. What is the surface area of this cylinder? Use $\frac{22}{7}$ for pi.

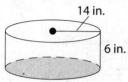

14 in.

6 in.

A 1760 in²

B 1176 in²

C 616 in²

D 3696 in²

9. What is the measure of angle ABC?

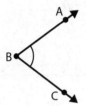

A

B

C

A 100°

B 360°

C 80°

D 270°

10. What is the volume of this triangular prism?

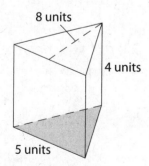

8 units

4 units

5 units

A 160 units³

B 72 units³

C 17 units³

D 80 units³

Name _____ Date _____

Placement Test

Circle the letter of the correct answer.

1. Which answer is the same as 8 + 6 + 9?

  A 8 tens and 6 ones

  B 6 tens and 9 ones

  (C 2 tens and 3 ones)

  D 3 tens and 8 ones

2. Look at the squares. How many groups of tens and ones are there?

  A 6 tens and 6 ones

  B 2 tens and 4 ones

  (C 3 tens and 6 ones)

  D 5 tens and 4 ones

3. What shape is missing from this pattern?

  A

  (B)

  C ◇

  D △

Name _____ Date _____

Circle the letter of the correct answer.

4. What number goes in the box to make the problem correct?

$\square + 14 = 19$

  A 6   B 8

  C 7   (D 5)

5. Look at the number line. If the rule continues, where will the arrow land on the next hop?

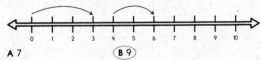

  A 7   (B 9)

  C 10   D 13

6. Which answer is an equation?

  A 9 = 3 = 6

  B 2 + 4 + 7

  C 7 − 2 + 7

  (D 8 + 3 = 11)

7. Pat saw 5 fish and 3 turtles in a pond. Then 2 frogs hopped into the pond. How many things in all were in the pond?

  (A 10)   B 8

  C 12   D 7

Name _____ Date _____

Circle the letter of the correct answer.

8. Which equation matches this problem?

There were 9 birds in a tree. All of them flew away. How many birds were left in the tree?

  A 9 + 1 = 10   (B 9 − 9 = 0)

  C 4 + 5 = 9   D 9 + 0 = 9

9. What is the perimeter of this shape?

  (A 15 cm)   B 5 cm

  C 8 cm   D 12 cm

10. Look at the graph. How many miles did Mom run?

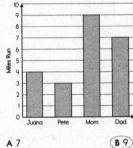

  A 7   (B 9)

  C 6   D 10

Placement Test

LEVEL **E**

Name _____ Date _____

Circle the letter of the correct answer.

1. Which answer is the same as 6 tens and 27 ones?

A 33
(B 87)
C 62
D 82

2. How many dimes are equal to 40 pennies?

A 10
B 400
C 14
(D 4)

3. Which answer is correct?

A 5 = 13
(B 4 < 9)
C 8 < 3
D 2 > 12

4. 9 + 7 = _____

A 9 − 7
B 9 + (7 + 9)
(C 7 + (6 + 3))
D 7 − (9 + 3)

LEVEL **E**

Name _____ Date _____

Circle the letter of the correct answer.

5. In which problem do you have to regroup?

(A 61 − 19)
B 54 − 42
C 38 − 25
D 76 − 30

6. Look at the pictograph. How many cars crossed the bridge on Wednesday?

Number of Cars Crossing Bridge

Monday	
Tuesday	
Wednesday	
Thursday	
Friday	
Saturday	
Sunday	

Key: 🚗 = 10 cars

A 17
B 24
C 53
(D 70)

7. What is the value of the digit 3 in the number 539?

A 13
(B 30)
C 39
D 300

LEVEL **E**

Name _____ Date _____

Circle the letter of the correct answer.

8. What number is missing in this function table?

IN	OUT
3	8
5	10
7	
15	20
23	28

A 33   B 14
C 29   (D 12)

9. A point on this graph is missing. The *x*-value of the point is 4. What is the *y*-value?

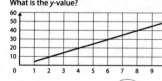

A 3   (B 20)
C 14   D 7

10. Which equation is correct?

A 5 + 9 = 14 − 5
B 5 + 5 = 9 + 9
(C 5 + 9 = 9 + 5)
D 9 + 5 = 14 − 9

Name _____ Date _____

Placement Test

Circle the letter of the correct answer.

1. Look at the number. Increase the value of the hundreds digit by 3. What would the new number be?

   24, 516

   A 24,519
   B 24,546
   C 27,216
   D 24,816 ⟨circled⟩

2. $936 - 478 =$
   A 458 ⟨circled⟩
   B 542
   C 452
   D 448

3. Which problem will have an answer less than 500?
   A $1162 - 584$
   B $267 + 294$
   C $1,427 - 986$ ⟨circled⟩
   D $143 + 406$

4. What is the value of b in this equation?

   $b + 9 = 15$

   A 7
   B 6 ⟨circled⟩
   C 3
   D 5

Circle the letter of the correct answer.

5. What is the range of a set of numbers?
   A the number that appears most often in the set
   B the highest number
   C the difference between the lowest and highest number ⟨circled⟩
   D the number in the set that appears only one time

6. $63 \div 9 =$
   A 6
   B 8
   C 4
   D 7 ⟨circled⟩

7. Which answer matches this problem?

   The temperature at four o'clock was 85°. By nine o'clock that night, it was 63°. How much did the temperature drop?

   A $85 + 63 =$
   B $85 - 63 =$ ⟨circled⟩
   C $85 + 4 - 63 =$
   D $9 - 4 =$

Circle the letter of the correct answer.

8. Look at the graph. In which game did Lee score more points than Chris?

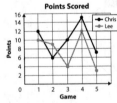

Points Scored

Chris — Lee

   A 1
   B 2 ⟨circled⟩
   C 3
   D 4

9. A driver delivered 214 soccer balls to a store. The next day, the driver brought 27 footballs and 39 basketballs to the same store. If 25 balls can fit on a shelf, how many shelves will be needed for all the balls?
   A 13
   B 15
   C 12 ⟨circled⟩
   D 19

10. What is the area of a room that is 8 feet wide and 12 feet long?
    A 96 square feet ⟨circled⟩
    B 20 square feet
    C 128 square feet
    D 82 square feet

Placement Test

Circle the letter of the correct answer.

1. Which figure has the smallest surface area?

A

B

C

D

2. $\frac{3}{4} \times \frac{2}{3} =$

A $\frac{3}{4}$

B $\frac{2}{3}$

C $\frac{1}{4}$

D $\frac{1}{2}$

Circle the letter of the correct answer.

3. $\frac{1}{4} + \frac{1}{3} + \frac{1}{12} =$

A $\frac{1}{4}$

B $\frac{5}{12}$

C $\frac{2}{3}$

D $\frac{2}{7}$

4. What is the median of the following set of numbers?

11 2 5 6 7 14 4

A 3

B 5

C 6

D 7

5. A board is 8 feet long. There are 12 inches in a foot. A worker cut the board into 6 pieces. How long was each piece of wood?

A 16 inches

B 14 inches

C 4 inches

D 48 inches

6. $6 \times (2 + 5) =$

A 42

B 13

C 17

D 67

Circle the letter of the correct answer.

7. Which answer is equivalent to $\frac{1}{4}$?

A $\frac{1}{8}$

B $\frac{4}{14}$

C $\frac{14}{28}$

D $\frac{8}{32}$

8. Which answer completes the equation?

$33 - 8 =$ ____ $+ 9$

A 25

B 17

C 16

D 34

9. There are three times more cows in a field than sheep. If there are 40 animals all together, how many of them are sheep?

A 12

B 10

C 4

D 20

10. Which answer is equal to 2^3?

A 2×3

B 3^2

C $2 \times 2 \times 2$

D $1 \times 2 \times 3$

Name _____ Date _____

Placement Test

Circle the letter of the correct answer.

1. In which answer are the numbers ordered from least to greatest?

  A $-2, -3, -5, 9, 13$

  B $-12, -4, 2, 7, 14$ *(circled)*

  C $2, -2, 3, -3, 5, -5$

  D $18, 12, 0, -6, -1$

2. The table below shows the pets owned by students in a class. What is the ratio of fish to cats?

PET	NUMBER
Fish	4
Birds	2
Dogs	9
Cats	7

  A 11   **B** 3

  C $\frac{4}{2}$ **D** $\frac{4}{7}$ *(circled)*

3. What is the volume of this figure?

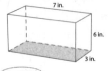

  A 126 in³ *(circled)*

  B 16 in³

  C 367 in³

  D 63 in³

Name _____ Date _____

Circle the letter of the correct answer.

4. $358\overline{)14962}$

  A 44

  B 42 R 9

  C 41 R 284 *(circled)*

  D 14 R 962

5. If you divided a circle into 3 equal angles, what would each angle measure?

  A 300 degrees

  B 120 degrees *(circled)*

  C 90 degrees

  D 63 degrees

6. What is 250% of 40?

  A 100 *(circled)*

  B 254

  C 425

  D 10

7. $\frac{2}{5} \div \frac{1}{8} =$

  A $2\frac{5}{8}$

  B $\frac{3}{13}$

  C $\frac{5}{16}$

  D $3\frac{1}{5}$ *(circled)*

Name _____ Date _____

Circle the letter of the correct answer.

8. What is the measure of the missing interior angle in this triangle?

  A 39 degrees   **B** 60 degrees *(circled)*

  C 40 degrees   **D** 120 degrees

9. What are the coordinates of point *c*?

  A $(2, -2)$ **B** $(-4, 1)$ *(circled)*

  C $(1, -3)$ **D** $(-4, -4)$

10. A box contains three balls of different colors. The colors are red, white, and blue. What is the probability of choosing the same color ball two times in a row?

  A $\frac{2}{3}$ **B** $\frac{1}{9}$

  C $\frac{1}{3}$ **D** $\frac{2}{27}$

Name _____ Date _____

LEVEL I

Placement Test

Circle the letter of the correct answer.

1. What is the least common multiple of these numbers?

3, 4, and 9

A 12
B 36
C 18
D 49

2. Which numbers are correctly ordered from least to greatest?

A $-\frac{5}{3}, -1, 0.05, 2\frac{1}{3}, \frac{9}{2}$

B $0, 1, \frac{2}{3}, -\frac{4}{2}, -0.7$

C $-0.5, -\frac{2}{3}, 1, \frac{5}{4}, 0.07$

D $\frac{1}{4}, -0.4, 2, -\frac{3}{5}, 4$

3. $63 - (-19) = $ _____

A 44
B 56
C 82
D 73

LEVEL I

Name _____ Date _____

Circle the letter of the correct answer.

4. $11\frac{1}{3} - 7\frac{3}{5} = $ _____

A $3\frac{11}{15}$

B $4\frac{1}{4}$

C 18

D $4\frac{1}{5}$

5. The regular price of a soccer ball is $28. It is on sale for 25% off. What is the sale price of the soccer ball?

A $21
B $3
C $7
D $25

6. $6.75 \div 0.25 = $ _____

A 6.5
B 4.5
C 13
D 27

7. Which of these proportions is not equal to the others?

A 1:3
B 4:5
C 15:45
D 6:18

Name _____ Date _____

LEVEL I

Circle the letter of the correct answer.

8. What is the surface area of this cylinder? Use $\frac{22}{7}$ for pi.

14 in.

6 in.

A 1760 in²   B 1176 in²
C 616 in²   D 3696 in²

9. What is the measure of angle ∧BC?

A
B
C

A 100°   B 360°
C 80°   D 270°

10. What is the volume of this triangular prism?

8 units

4 units

5 units

A 160 units³   B 72 units³
C 17 units³   **D 80 units³**